6th Grade Math
Volume 6

© 2013 OnBoard Academics, Inc
Newburyport, MA 01950
800-596-3175
www.onboardacademics.com

Table of Contents

Metric Measure for Mass & Capacity

Key Vocabulary

metric units

mass

capacity

Measuring length: mm and cm.

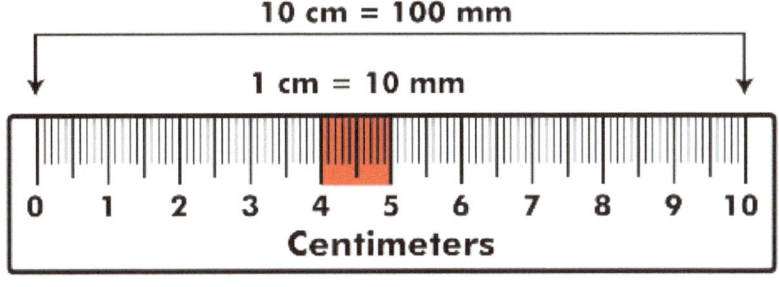

3 cm = [ ] mm

7.3 cm = [ ] mm

111 mm = [ ] cm

Measuring length: m, cm and mm.

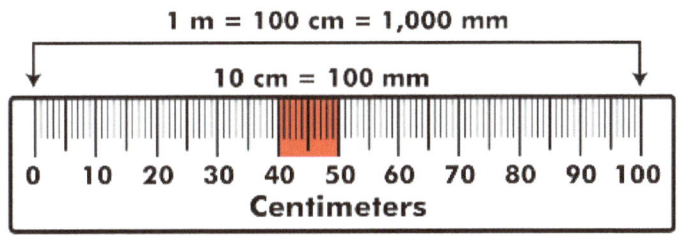

0.5 m = [ ] cm

0.5 m = [ ] mm

150 cm = [ ] m

Measuring length: km, m, cm and mm.

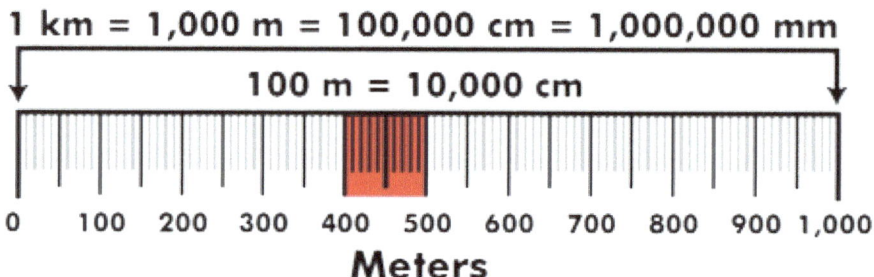

1 km = 1,000 m = 100,000 cm = 1,000,000 mm

100 m = 10,000 cm

0   100   200   300   400   500   600   700   800   900   1,000

Meters

1.5 km = ☐ m

1.5 km = ☐ cm

600 m = ☐ km

**Write the appropriate metric measurement from the suggestions given.
Measure each items.**

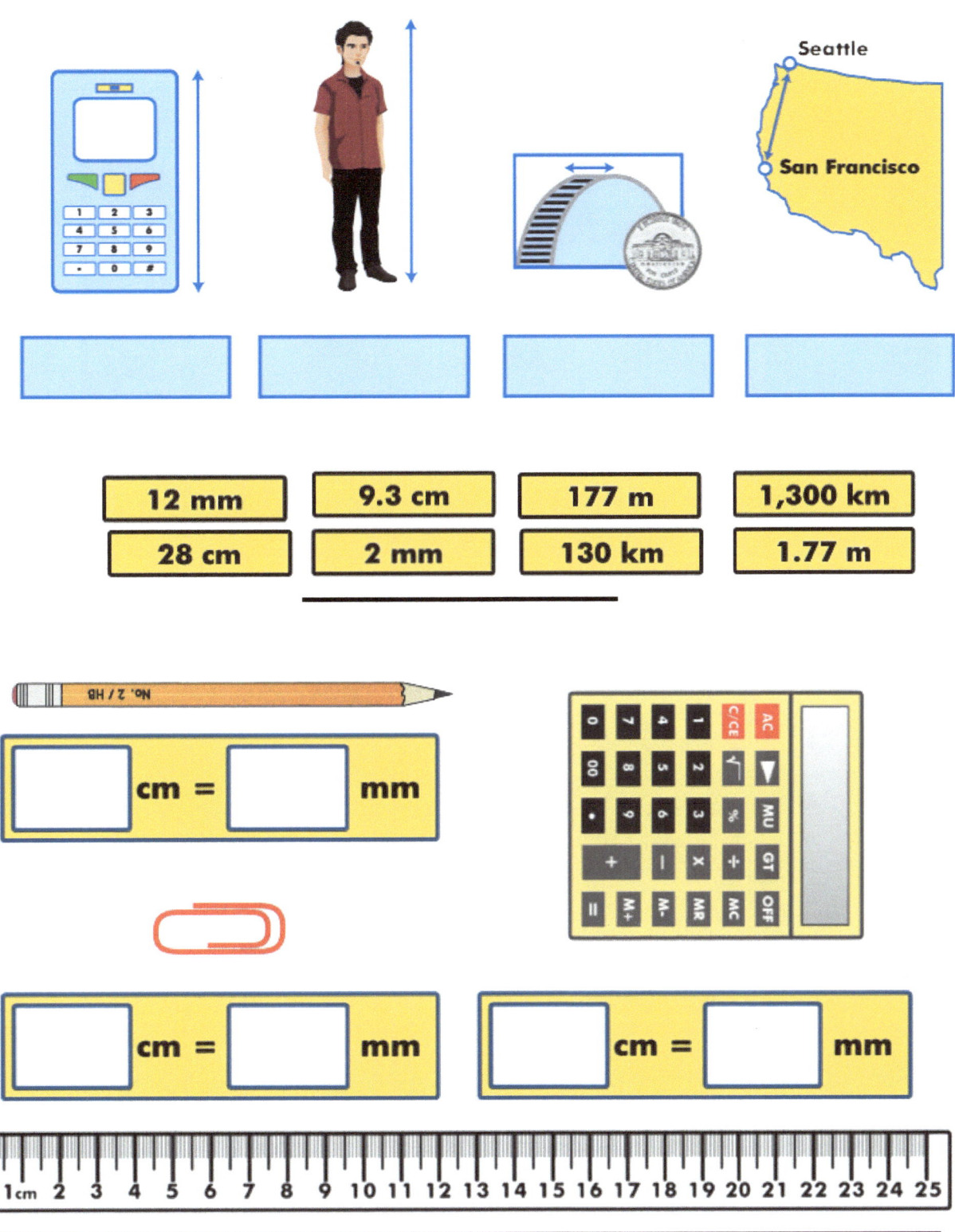

| 12 mm | 9.3 cm | 177 m | 1,300 km |
| 28 cm | 2 mm | 130 km | 1.77 m |

cm = mm

cm = mm   cm = mm

Convert metric units of mass.

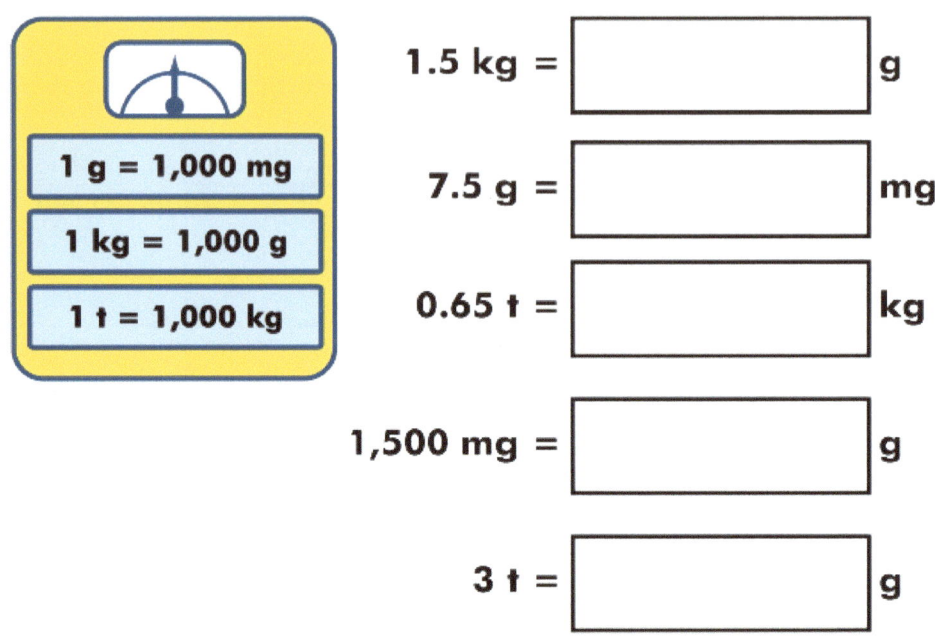

1 g = 1,000 mg

1 kg = 1,000 g

1 t = 1,000 kg

1.5 kg = ☐ g

7.5 g = ☐ mg

0.65 t = ☐ kg

1,500 mg = ☐ g

3 t = ☐ g

What is the most appropriate mass for each item..
Choose from the suggestions given.

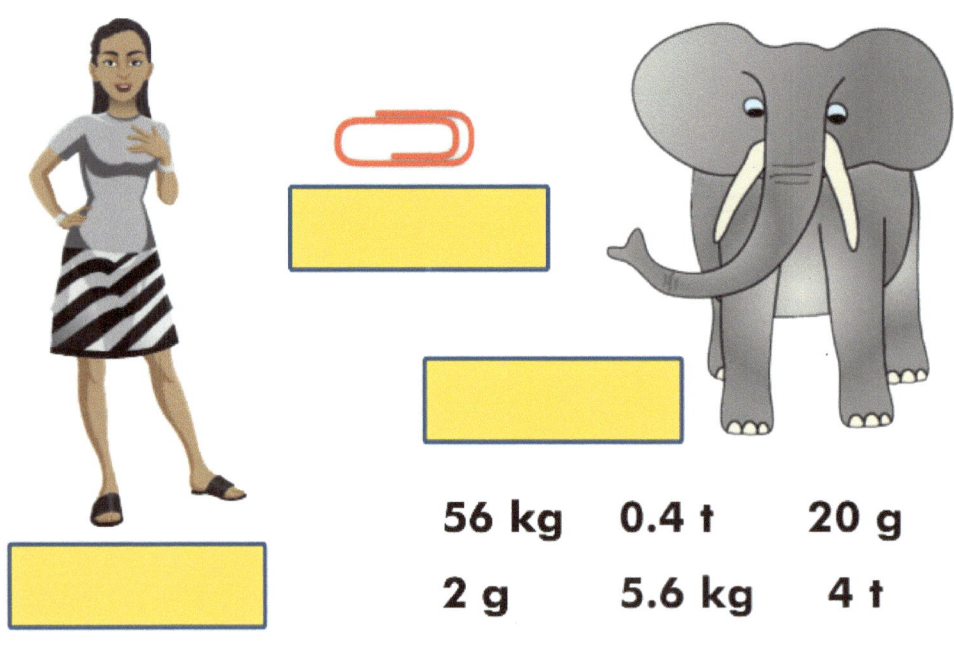

56 kg   0.4 t   20 g

2 g   5.6 kg   4 t

Estimate the capacity of each of these containers in ml or l.

Name_____

Metric Measure for Mass and Capacity Quiz

 **True or false, metric measures for capacity include liters, centimeters and milliliters?**

 **Which of the following statements is *not* correct?**

A 1 m = 1,000 mm

B 1 l = 1,000 ml

C 1 km = 1,000 cm

D 0.01 km = 1,000 cm

3 **Write 5.5 liters in milliliters.**

4 **Write 12,000 m in km.**

Area of Irregular Figures

Key Vocabulary

area

irregular figures

Study the illustrations below to discover strategies for finding the area of an irregular figure.

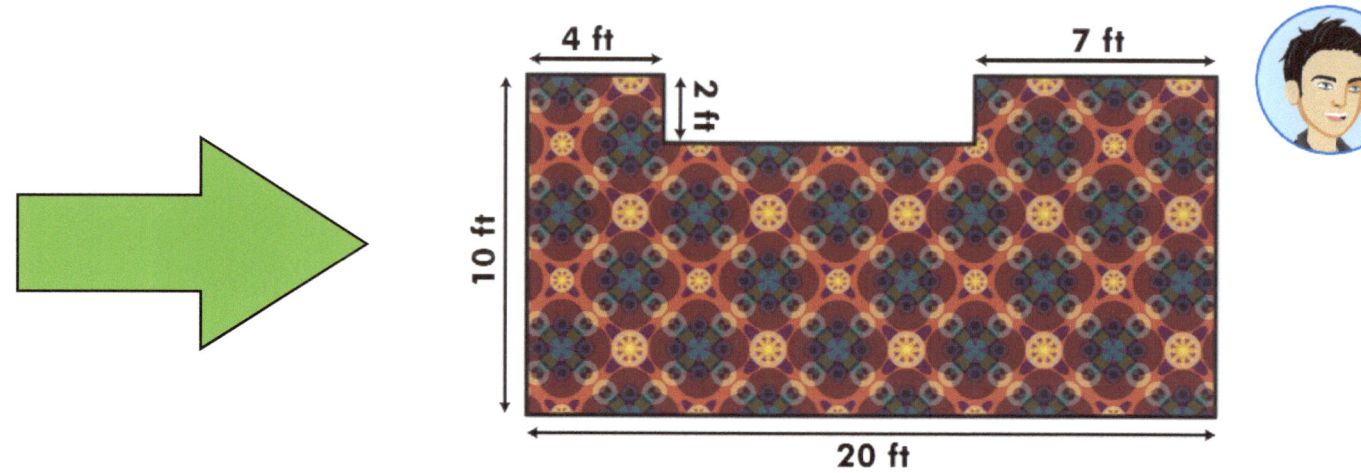

Ben's parents have reluctantly agreed to let him host a party, but they've insisted that he cover the carpet with a protective plastic sheet. What size sheet does he need?

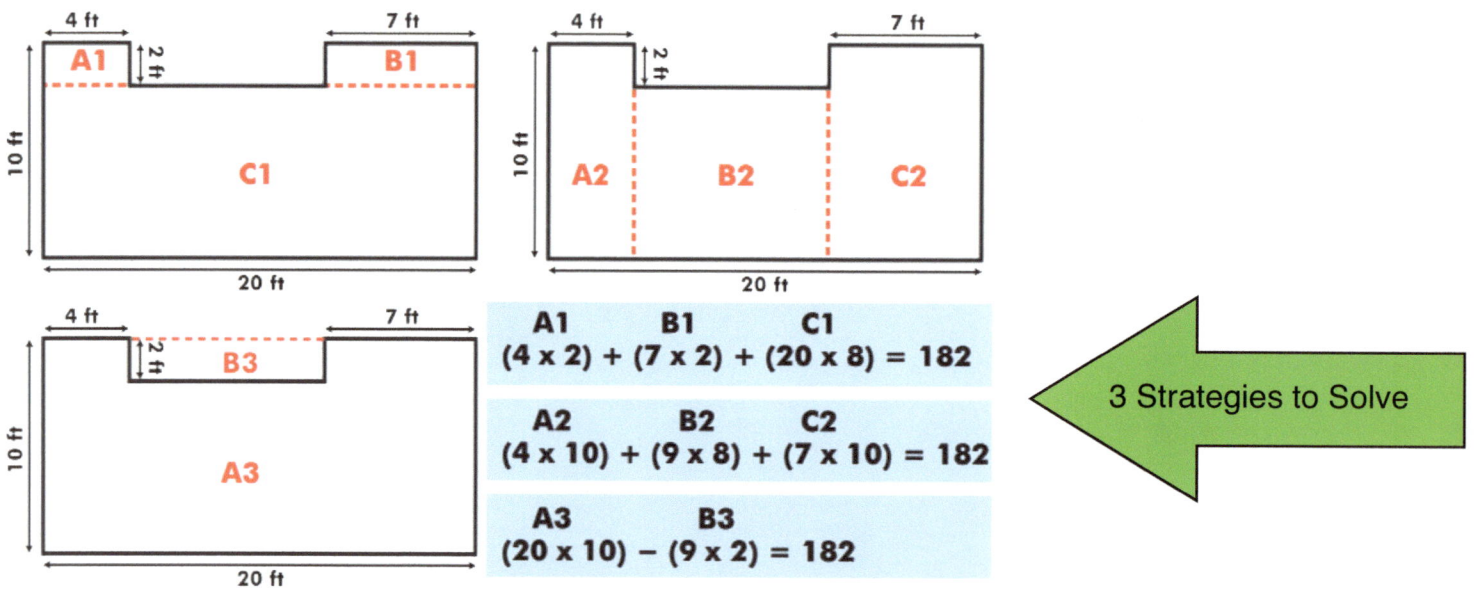

A1   B1   C1
(4 x 2) + (7 x 2) + (20 x 8) = 182

A2   B2   C2
(4 x 10) + (9 x 8) + (7 x 10) = 182

A3   B3
(20 x 10) − (9 x 2) = 182

3 Strategies to Solve

Ben needs 182 sq ft of sheeting to cover the carpet.

Practice dividing irregular shapes into familiar shapes.

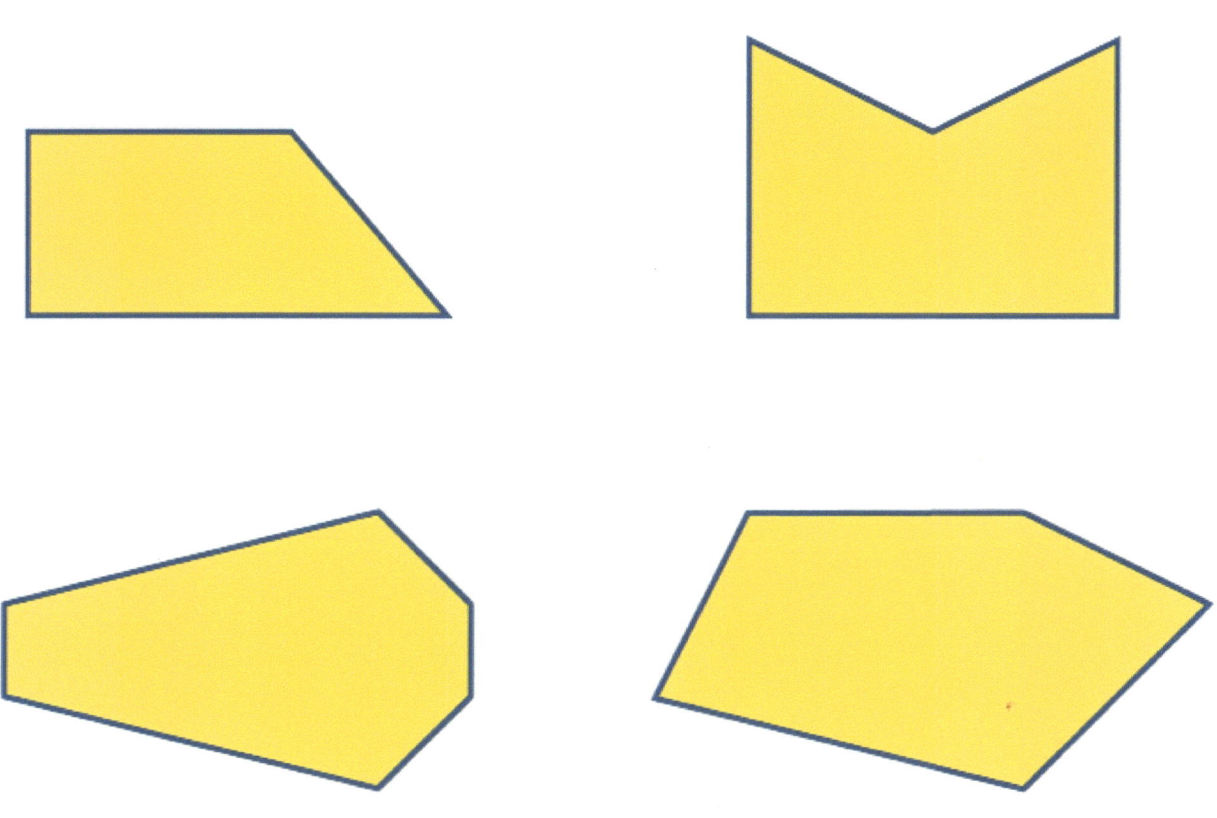

Find the area of these irregular shapes.

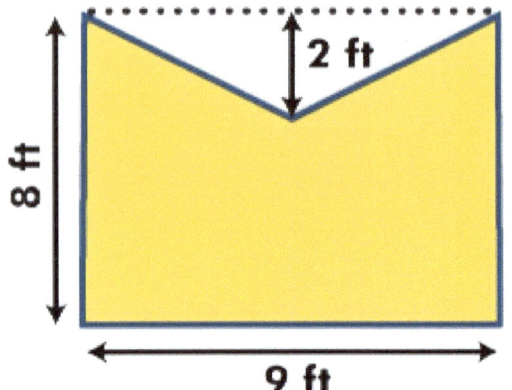

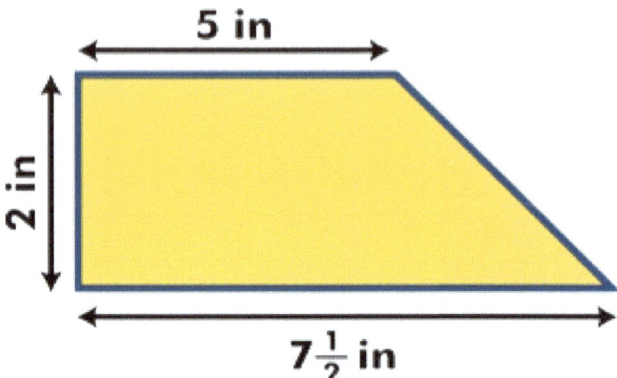

Find the area of this irregular hexagon.

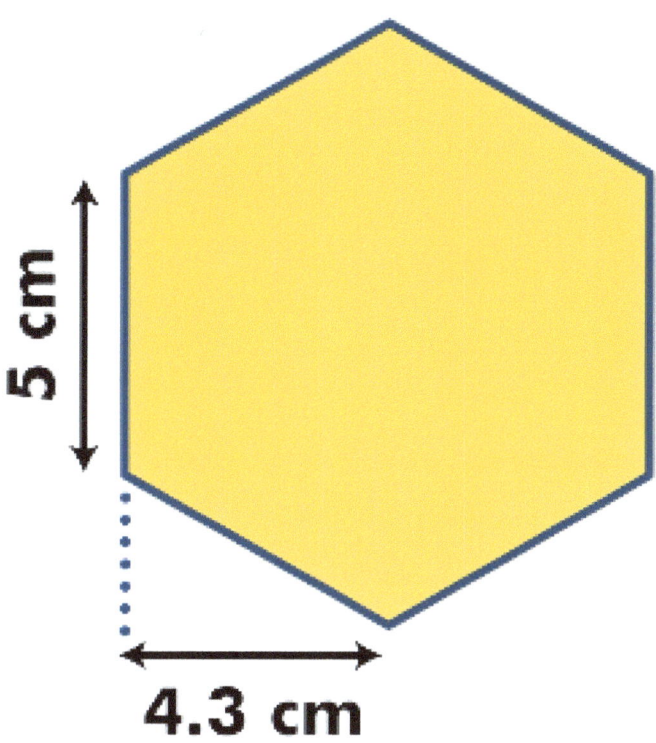

Who has the largest deck?

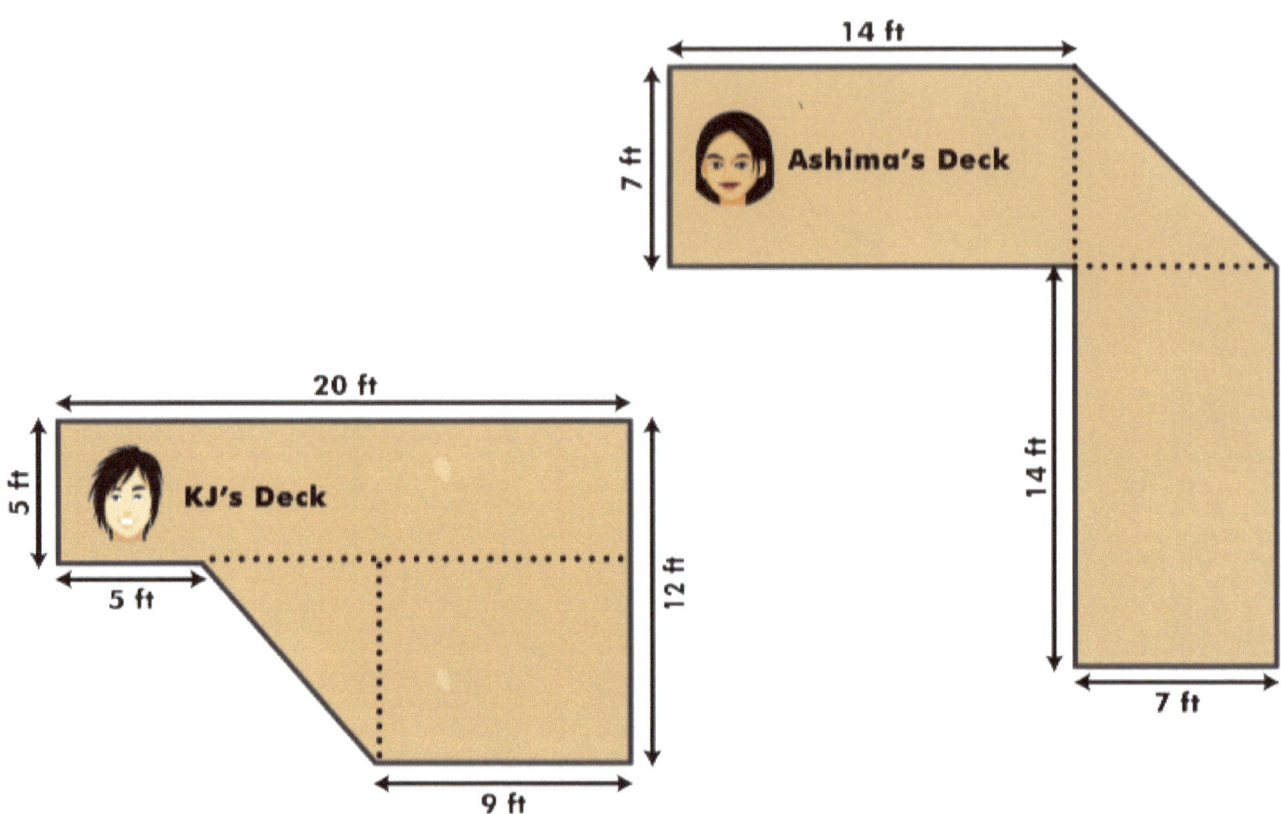

Name_____

Area of Irregular Figures Quiz

1   True or false? The area of the arrowhead (the triangle) in Figure 1 is 12 sq units.

2   The total area of Figure 1 is:

    A   38 sq units

    B   40 sq units

    C   44 sq units

    D   32 sq units

3   What is the area of Figure 2 in sq units?

4   What is the value of x in Figure 3?

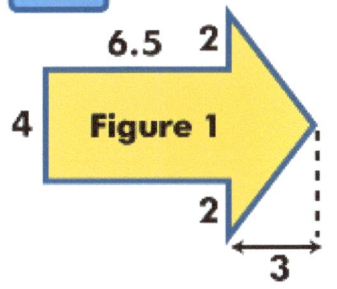

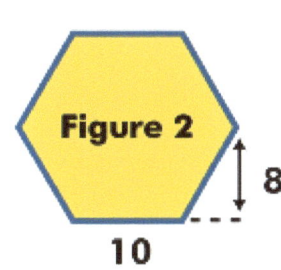

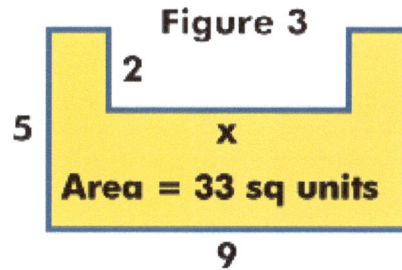

Nets and Surface Area

Key Vocabulary

surface area

net

prism

cube

rectangular prism

How many faces does a cube have?

Take it apart to solve.

Finish the nets.

A net is a pattern that you can cut and fold to make a model of a solid shape. In this case we want to be able to make a cube. Complete the nets by shading in the additional squares to make a cube.

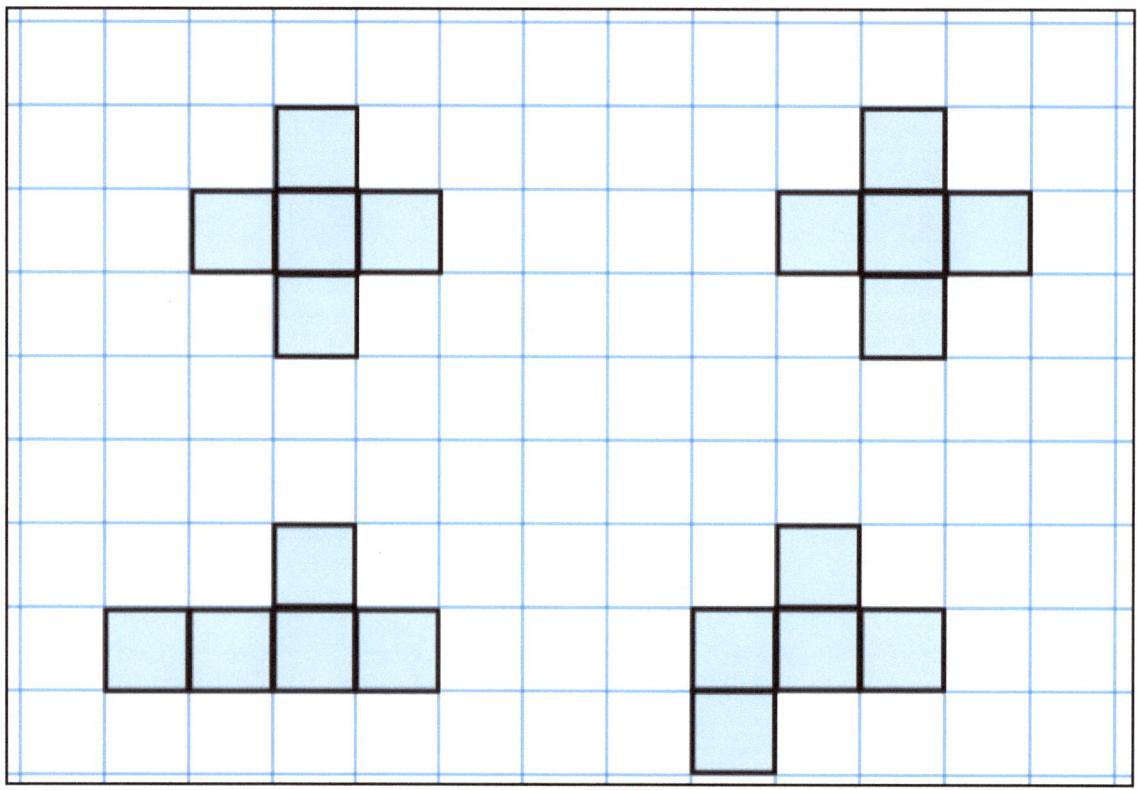

Which of these nets will make a cube.
Place a Y in the box for yes and an N for no.

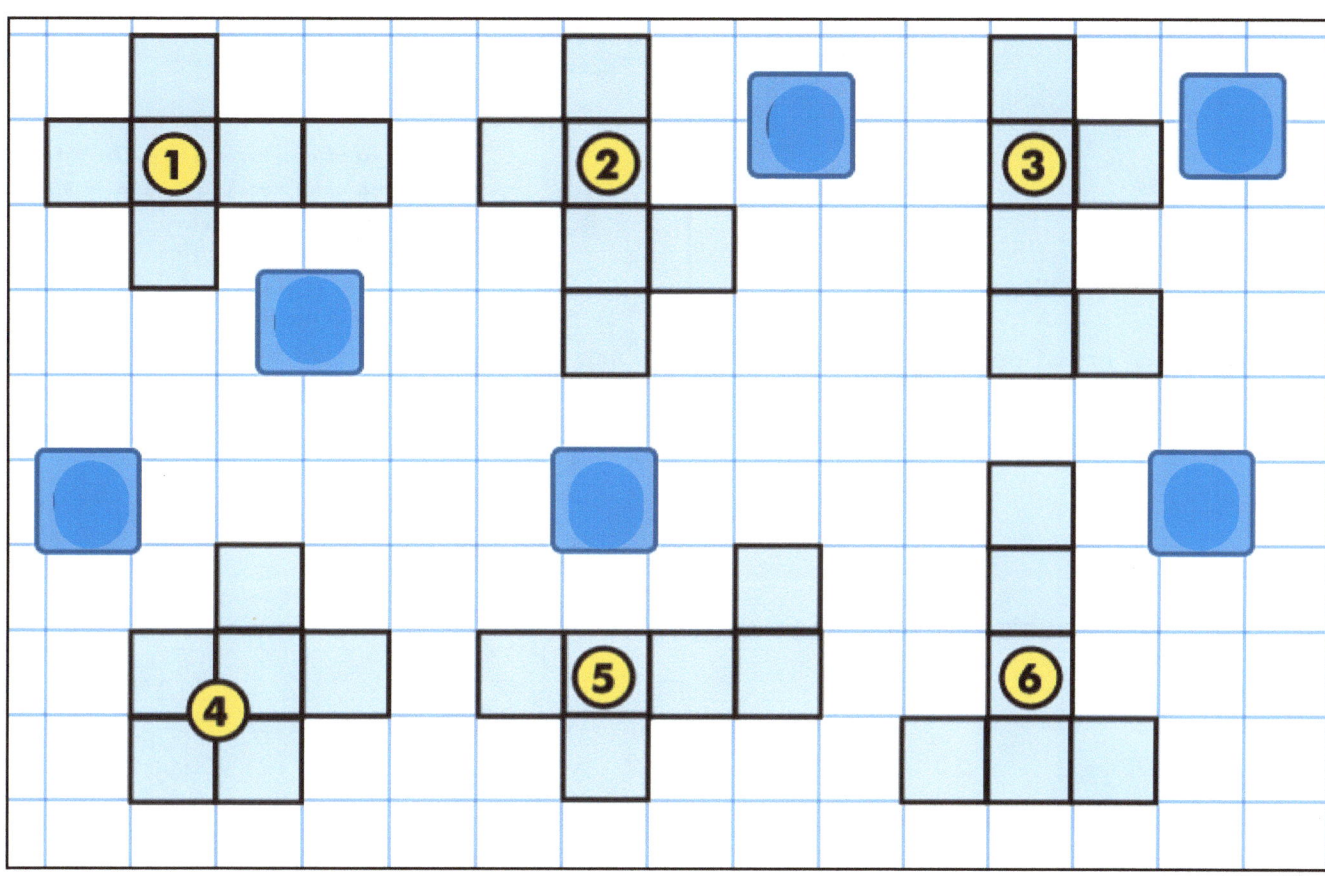

The surface are of a cube.

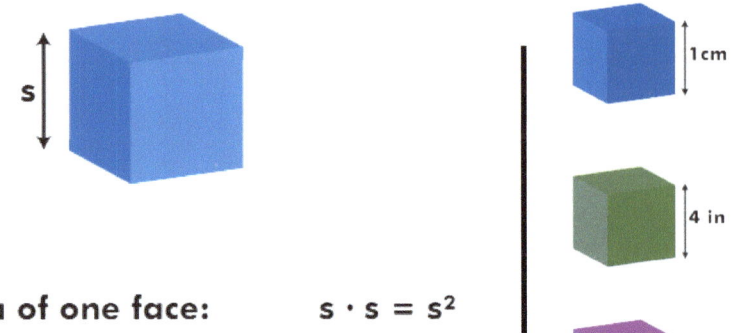

Area of one face: $s \cdot s = s^2$

Surface area of cube: $6s^2$

Area of one face: 1 cm x 1cm = 1 cm²
Area of six faces: 6 x 1 cm² = 6 cm²
Surface area = 6 cm²

Area of one face: 4 in x 4 in = 16 in²
Area of six faces: 6 x 16 in² = 96 in²
Surface area = 96 in²

Area of one face: 1.5 in x 1.5 in = 2.25 in²
Area of six faces: 6 x 2.25 in² = 13.5 in²
Surface area = 13.5 in²

What is the surface are of these cubes?

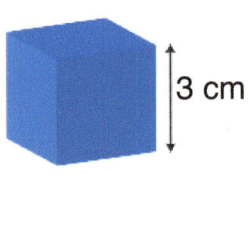
3 cm

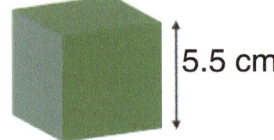
5.5 cm

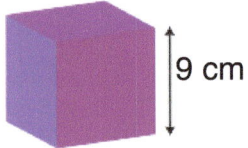
9 cm

The Net of a Rectangular Prism

Study the illustrations below to discover how to draw the net of a rectangular prism.

Surface area of a rectangular prism.

Draw a net for this figure.

Calculate the surface area. _____

Find the surface area for this rectangular prism using what you have learned.

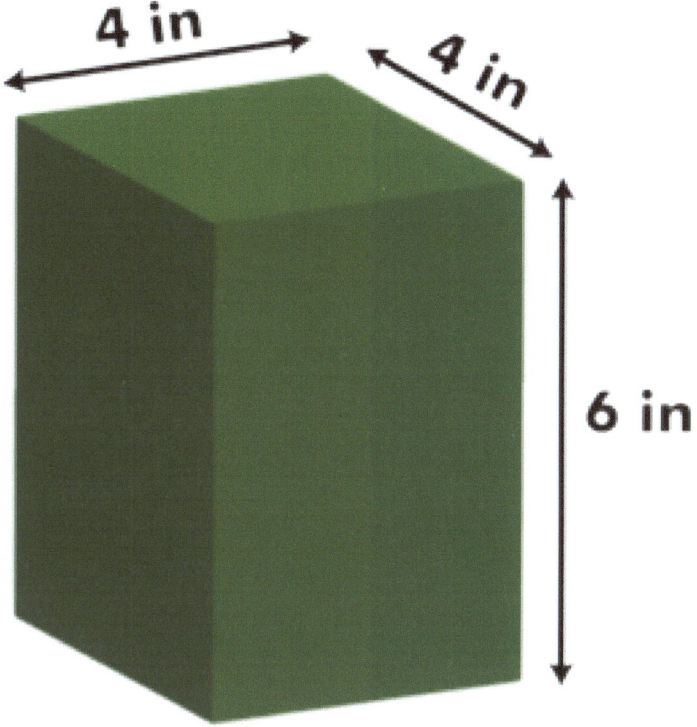

Name_____

Nets and Surface Area Quiz

1 True or false? Figure 1 is the net of a rectangular prism.

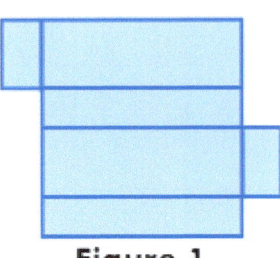

Figure 1

2 The length of the side of a cube is 5 cm. The surface area of the cube is:

A 25 cm^2

B 125 cm^2

C 625 cm^2

D 150 cm^2

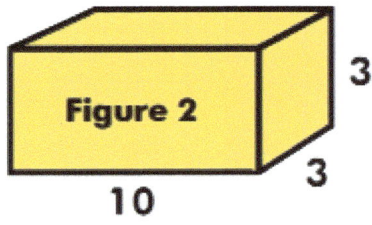

Figure 2

3

3

10

3 What is the surface area of Figure 2 in square units?

Volume of a Rectangle

Key Vocabulary

volume

cube

cube units

Rectangular Prism Volume

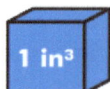

How many cubes in the first layer?	16
Number of layers?	3
Total number of cubes?	3 x 16 = 48
Volume of rectangular prism?	48 in³

Complete the table and find the volume of these prisms.

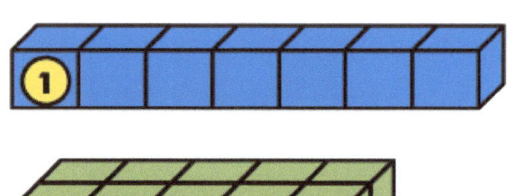

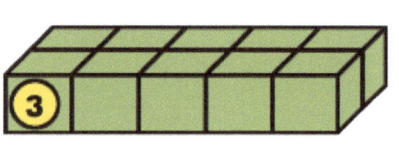

	Length	Width	Height	Volume
1	in	in	in	in³
2	in	in	in	in³
3	in	in	in	in³

What is the volume of these prisms?

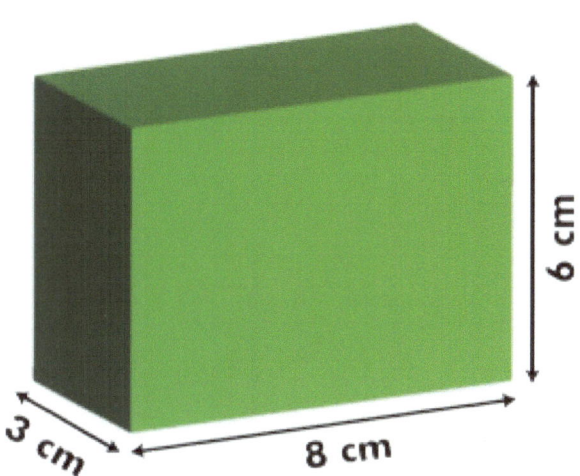

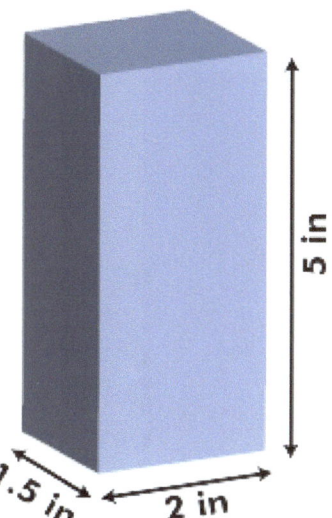

Good Boy Treats Packaging
How many of the small packages of Good Boy Treats fit in the large crate?

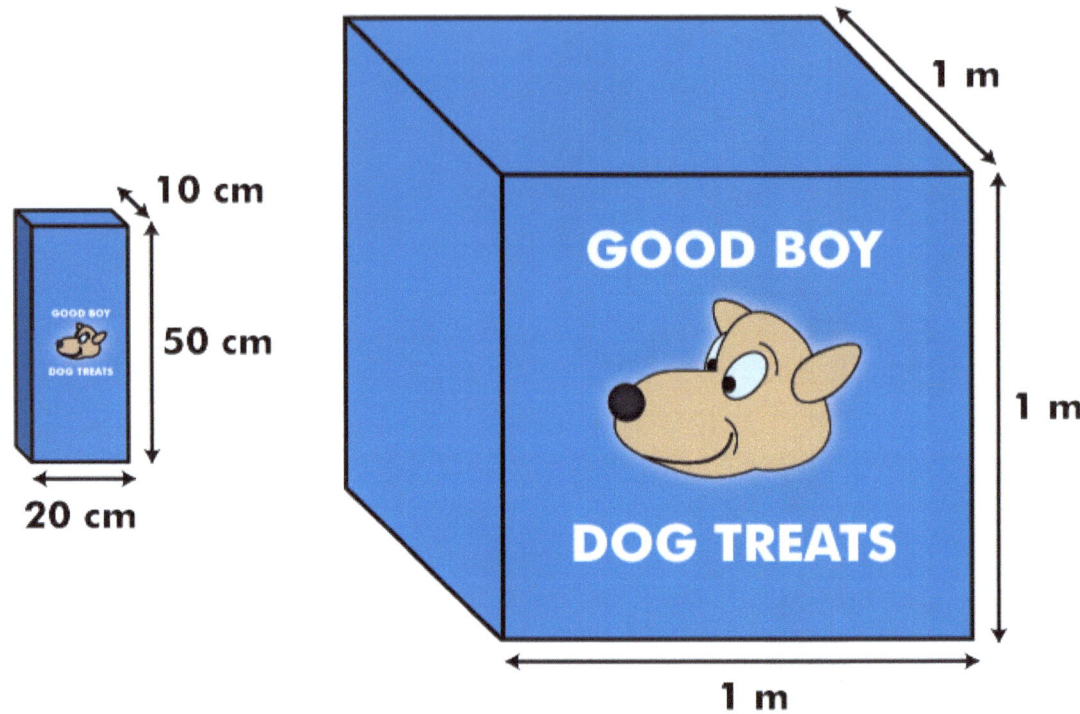

Try to model the problem to help to solve.

Can you create three different size pieces of luggage that are compliant with Econo-Air's baggage policy?

ECONO-AIR CARRY-ON POLICY
The length + width + height of your carry-on luggage must not exceed 45 inches.

	Length	Width	Height	Volume
1				
2				
3				

Name_____

Volume of a Rectangle Quiz

1 True or false? Volume is measured in square units.

2 The formula for the volume of a rectangular prism is:

A l x w x h

B 2l x 2w x 2h

C l + w + h

D (l + w + h)³

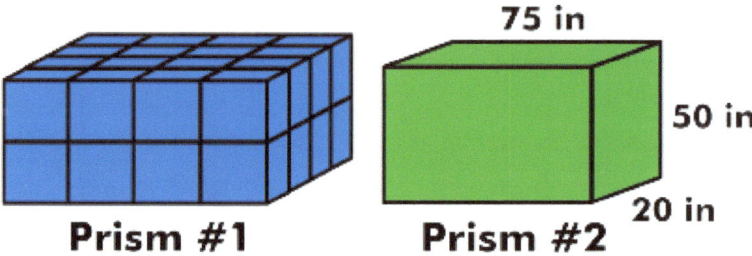

Prism #1   Prism #2

75 in

50 in

20 in

3 How many cubes are there in prism #1?

4 What is the volume of prism #2 (in cubic inches)?

Newburyport, MA 01950

1-800-596-3175

OnBoard Academics employs teachers to make lessons for teachers! We create and publish a wide range of aligned lessons in math, science and ELA for use on most EdTech devices including whiteboard, tablets, computers and pdfs for printing.

All of our lessons are aligned to the common core, the Next Generation Science Standards and all state standards.

If you like our products please visit our website for information on individual lessons, teachers licenses, building licenses, district licenses and subscriptions.

Thank you for using OnBoard Academic products.